THE ORIGIN OF LIFE

(Original French title:
Aux Origines de la Vie)

by

Bernard Hagene and Charles Lenay

Illustrated by Michel Loppé

Translated from the French by
Albert V. Carozzi and Marguerite Carozzi

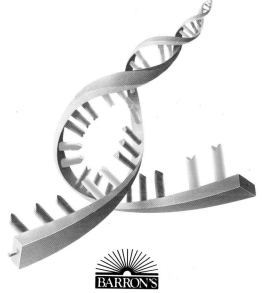

BARRON'S

New York • London • Toronto • Sydney

First English language edition published in 1987 by
Barron's Educational Series, Inc.

© 1986 Hachette — Fondation Diderot/La Nouvelle Encyclopédie,
Paris, France

The title of the French edition is *Aux Origines de la Vie.*

All inquiries should be addressed to:
Barron's Educational Series, Inc.
250 Wireless Boulevard
Hauppauge, New York 11788

International Standard Book No. 0-8120-3841-X

PRINTED IN FRANCE
789   9687   987654321

Contents

The Mystery of the Origin

"The first man created by Mwari, the creator, was called Moon. He dwelled at the bottom of the water. One day he wanted to live on the earth, which was then completely arid. The creator gave him a spouse, Morning Star. The next day Morning Star gave birth to grass, bushes, and trees. Two years later, Moon took another spouse, Evening Star. The next day she gave birth to chicken, sheep, and goats.''

Bantu tale, southern Africa

This supernatural tale of the origin of living beings is only one among hundreds that could be quoted. On each continent, in every civilization, a ''history'' explains the beginning of what can be seen in the sky and on the earth. Each ''history'' is an answer imagined by people in search of an explanation. For a very long time, these explanations almost always relied upon sudden events associating the natural and the supernatural.

After the eighteenth century, these explanations were less and less accepted, at least in the Western world. Whereas the questions asked have remained the same—Where do heavenly bodies, living beings, men, come from?—the methods of interrogating the world have evolved. Scientific methods were invented to observe and compare, to experiment—and particularly, to reject unacceptable hypotheses.

Origin *The origin of life is still debated at the beginning of the twenty-first century. This problem, so difficult to resolve, requires investigations in all fields, from the immensity of space to the infinite smallness of matter.*

For the matter under discussion—the origin of life—a leading principle had to be found. This was a very long and arduous undertaking.

A few ideas appeared at the end of the nineteenth century. Scientific progress between 1920 and 1930 resulted in the development of better research techniques. After 1950, research became more intensive. Various branches of science brought their own approaches to try to unravel the mystery.

Among the more or less successfully explored paths, amid the interplay of argument and experiment that divide and unite scientists, fruitful ways were found, while some enigmas remained. This is what we will discuss in the pages that follow.

The debate about the origin of life remains *alive*.

The Mystery of the First Billion Years

The origin of life is a gap in the history of the universe from its beginning until today, similar to a missing chapter in a novel.

Astrophysicists tell us how galaxies, stars, and planets, and in particular the sun and the earth, were formed. Geologists follow in their path and reconstruct the history of the earth by the study of its rocks. They think that the earth is 4.6 billion years old. But its rocks have been deeply reshaped, and the most ancient rocks at the surface of the earth's crust are only 3.8 billion years old. Paleontologists study successive forms of life on earth, and find traces of life in fossils.

What happened during the first billion years on earth? It is during that time that the first forms of life must have appeared.

Biologists and chemists know many things about life as observed today, about the way it is organized and how it functions.

What is Life?

Some people think that before searching for the origin of life, one should know what life is. No precise answer is available. Biologists disagree on how to define what is alive and what is not.

The question is obviously irrelevant for a large animal or a plant. But for a

Between the origin of the universe and today, life appeared. This event happened some time between the formation of the earth (4.6 billion years) and the first occurrence of traces of organic matter in the oldest known rocks (3.8 billion years).

very small microbe, or a virus, how does one know whether or not it is alive?

Self-Preservation

The most important characteristic of a living being is its capacity for self-preservation. Indeed, a living organism is first of all something capable of surviving most odds. Although external conditions change continuously, the organism resists and survives, except under extremely harsh conditions, when life becomes impossible, and death ensues.

Self-Regulation

Man, a large mammal, has an internal temperature that always remains at 37°C (98.6°F). In warm weather, automatic cooling processes operate. Small blood vessels expand, and man perspires. Sweat evaporates and prevents an increase in the internal temperature. In cold weather, the organism becomes more active, the body shivers slightly and warms up.

This is one process of self-regulation. Other examples involve respiration and nutrition. The amount of oxygen or the assimilated amount of various kinds of nourishing sub-

Self-preservation. *All year long—summer and winter, in the sun or under the snow, with or without leaves—a tree remains alive. A living being has the capacity for self-preservation. Death occurs when the environmental conditions exceed this capacity: long winters, forest fires, etc.*

stances are kept at a constant level in the internal environment of a living organism.

If self-regulation is disrupted, the organism can no longer withstand environmental variations and dies.

Self-Organization

Sometimes the assaults of the external environment may be more coarse: a shock, for example, or a cut. If the wound is a minor one, the living organism repairs itself. It reconstructs the damaged portion by forming a scar. Some organisms are able to reconstitute themselves to a greater degree: a torn-off tail of a lizard grows back. This is self-organization.

If a repair is not possible because of a major wound or because the organism is too weak, death occurs.

Self-organization is organized growth: we build ourselves. A living organ-

ism can take energy and matter from various foodstuffs to build itself and grow. For humans, this self-organization begins with the very first small cell in the maternal womb. This cell multiplies billions of times and forms gradually an embryo, which grows and becomes extremely complex. After birth, self-organization continues until an adult man or woman is formed.

Self-Reproduction

A pigeon, a whale, a human being—all die some day. But the pigeon species, the whale species, and the

Self-regulation. *Certain living organisms regulate their internal temperature. For instance, the polar bear, protected by its fur and a thick layer of fat, expends much energy to keep warm. The dromedary, on the other hand, strives to keep cool: its temperature cannot go beyond 41° C (105.8° F). Its hump, full of fat, acts as an energy reserve.*

human species remain living species. These species are preserved because the offspring of pigeons are pigeons, the offspring of whales are whales, and the offspring of human beings are human beings. This is self-reproduction.

One speaks of self-reproduction because a living being does not reproduce in the same way as television sets are manufactured. Machines that make televisions in a factory are not televisions themselves. On the contrary, living individuals are capable of making individuals like themselves. This is the marvelous self-reproduction capacity of living beings.

The Cell

Observing living beings under the microscope shows that they consist of cells. A human body has billions of cells, but certain

Self-reproduction of species. *How great is the self-reproduction capacity of rabbits! A pair can have three to four litters of ten offspring per year. Under optimum conditions, there could theoretically be more than one million descendants within five years!*

very small beings, such as bacteria, are made of a single cell. A cell is, therefore, already life. To understand the origin of life, it would be very helpful to understand the origin of the cell.

What Methods Should Be Used to Understand the Origin of Life?

How do scientists proceed to fill the gap in our knowledge? Two paths of investigation are open to them:

—To proceed from the simplest to the most complex: to visualize how the starting blocks can organize themselves into shapes that resemble life more and more closely.

Each hypothesis should be verified by laboratory experiments. Each successful experiment would enable the scientist to visualize the next step toward a more complex organization.

—To proceed from the most complex to the most primitive: to observe with care increasingly older traces of fossil life and search for increasingly rudimentary beings, looking for forms that may have preceded life.

To understand the origin of life would be to succeed in linking these two paths of investigation, thus joining together imagination, experimentation, and observation.

From Spontaneous Generation to Pasteur

The modern approach to the study of the origin of life began in the years 1920 to 1930.

For a long time, it was believed that living beings could suddenly appear inside inert matter—worms from cheese or wood, insects from mud, for example. This was called "spontaneous generation." A remarkable example was given by the Belgian chemist Jan Baptiste van Helmont in the seventeenth century: "Mice can be produced simply by leaving wheat and dirty linen in a box. It is proven by experience!"

But at the end of the seventeenth century, several scientists began to question this way of thinking. The Italian biologist Francesco Redi showed that maggots would not appear in a piece of meat in which flies were prevented from laying eggs. For him, life could only come from life. Then, with the help of the microscope, the world of microscopic beings was discovered.

Later on it was thought again that living matter could only be born spontaneously in the liquid in which it is found. This idea, although strongly disputed, was still widespread until about 1860.

This is when Louis Pasteur came in. The French scientist had studied different types of microscopic life for a long time. He knew well the precautionary measures necessary to complete an experiment and keep it free from contamination. He showed that microscopic life occurs everywhere and in huge numbers: in the air, on the walls of laboratory vessels, on hands.

Pasteur's work and that of others proved that life always originates from preexisting life and that spontaneous generation does not exist.

Life Scientists

A. Oparin

Miller, Stanley L. *U.S. chemist who demonstrated the possibility of spontaneous formation of amino acids under the conditions assumed to prevail on the primitive earth.*

Oparin, Alexandr I. *(1894–1980). Soviet scientist, the father of modern ideas on the origin of life. As early as 1924 he proposed the hypothesis that simple life forms appeared in an already complex environment charged with chemical energy.*

Haldane, John B. S. *(1892–1964). English mathematician and biologist. At the same time as Oparin, he proposed the hypothesis that life appeared gradually in what he called "the primitive soup."*

Monod, Théodore. *French biologist and naturalist. While exploring the Sahara, he discovered in Mauritania rocks one billion years old.*

Boureau, Édouard. *French paleobotanist. He studied the large spheroids found in Mauritania. He collaborated with Marcel Locquin in developing a staining technique for fossils that reveals their organic matter.*

Fox, Sidney. *U.S. scientist who in the 1960s discovered his famous microspheres and a process of proteinoid formation by heat.*

Oró, John. *U.S. scientist who, continuing the work started by Miller, demonstrated the spontaneous synthesis of nucleic bases.*

Pflug, Hans Dieter. *West German scientist who discovered the first organic molecules in rocks (3.6 billion years old).*

13

The Primitive Earth

Nothing is more commonplace than the matter of which living beings are made. In this matter, one finds again and again the same chemical elements— about twenty: first, carbon, hydrogen, oxygen, and nitrogen; then, some chlorine, calcium, sodium, potassium, magnesium, sulfur, and iron; and finally, a very small amount of a few other elements.
This list is not unusual. These elements occur in all the materials forming the earth, and, in fact, these same elements occur everywhere in the universe.

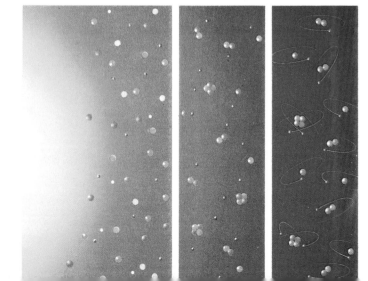

Yet in another way, this commonplace list is very extraordinary: it shows that life is an intrinsic part of the universe and that living organisms do not include elements other than those already existing and common in the inert matter that forms our planet with its minerals, oceans, and atmosphere.

Two important questions arise:

—Where did the elements that occur in living matter come from?

—How was the planet earth born? More precisely, how did the "primitive earth" on which life appeared look?

From the Big Bang to the present:

1. One billionth of a second after the initial explosion—the Big Bang—elementary particles and antiparticles were made and destroyed in an ocean of light. After one hundredth of a second, only neutrons, protons, and electrons remained.

2. A time of three to twenty-five minutes was necessary for the temperature to drop to one billion degrees, so that combinations of neutrons and protons formed the first stable atomic nuclei.

3. At a temperature of about 3000°, nuclei were surrounded by electrons and formed atoms.

4. Finally, the era of stars and galaxies began. Our solar system appeared ten billion years after the Big Bang.

Stars as Crucibles of Chemical Elements

Our universe began fifteen billion years ago in a kind of explosion, the Big Bang, which liberated an incredible amount of energy. Radiation and particles surged and continuously changed into one another, while the universe expanded and the temperature of the original blaze dropped rapidly. Only after a few minutes, when the temperature was down to about one billion degrees, could nuclei of stable atoms persist.

Thus was formed matter consisting of three parts hydrogen and one part helium. Huge clouds of these elements moved off

A supernova: the Crab nebula. In 1054, Chinese astronomers observed the appearance of a new star. At the beginning, it was so brilliant that it could be seen in full daylight. Its luminosity gradually decreased and, after a few months, disappeared. In fact, it was an old star that had just exploded, scattering its matter in space: a supernova. In the envelope that continues to expand and that can still be observed today with the telescope, elements exist that were gradually produced in the center of the primitive star.

into space. In certain places, they agglomerated and contracted at the same time, while their temperature increased. When a cloud was sufficiently compressed, the temperature was so high that nuclear reactions similar to those in hydrogen bombs were initiated: a star began to shine!

If such a star has a large mass, its life will be tempestuous. It will consume "fuel"—hydrogen—and produce "ashes"—helium. After a new brutal contraction, it will "burn" helium and produce new ashes: carbon. From contraction to contraction, it will produce a series of elements: oxygen, silicon, iron. Finally, the iron center will

Formation of the solar system: *In a cloud of whirling matter, most of the gases collected to form the sun. Remaining dust particles were attracted to one another and became glued together, first forming flakes, then grains. They continued to rotate around the just born sun. Colliding with other grains, they either broke apart or stuck together to form larger and larger bodies. After a hundred million years, a series of planets was formed.*

collapse, producing an enormous explosion: a supernova will appear in the sky. The peripheral layers of the star will be projected into space, and nitrogen, copper, magnesium, and many other elements will be formed.

This chronology is very important for our history. The space between stars is

Immediately after its formation, the surface of a planet such as the earth was riddled with craters, sometimes huge, formed by falling meteorites of variable size. Each impact released heat. The earth was therefore very hot. Volcanic activity was intense. The surface of the earth remained more or less liquid during some hundred million years, before its crust cooled and hardened. At that time, torrents of water pouring from the sky began to form the oceans. The atmosphere was streaked by lightning. Nothing shielded the earth from the ultraviolet radiation of the sun, and radioactivity was intense.

19

seeded by different atoms, which in turn form interstellar clouds. From one of these clouds the sun and its planets were born. Stars are therefore the machines that produced the atoms necessary for the formation of the earth and of living beings.

The Birth of the Earth

Dust particles and gases formed from exploded stars expanded in space as gigantic clouds. Within one of these clouds a new regrouping began. While whirling, almost all the matter reassembled to form a new star: the sun. But the remainder transformed

The Blue Planet

Astronauts, direct observers of the earth, all told about their wonder when gazing at the "blue planet." The nine planets that circle the sun were formed simultaneously from the same cloud of gases and dust particles. They are nevertheless very different. The earth, with two-thirds of its surface covered by oceans, is exceptional. Without water, life would not have begun or evoloved into its present forms. The presence of water and its continuous existence during billions of years are related to the evolution of the primitive atmospheric layer that surrounded the earth. But the primitive atmospheres of Venus or Mars were unquestionably similar to that of the earth. Why then such differences? Scientists undertook very elaborate calculations to answer this question. Even if these calculations are somewhat approximate, their general conclusions are probably correct. They show that if the earth had been closer to its star—if it had circled at a distance of 140 million kilometers (87 million

itself into a few large planets, one of which is the earth, and into a great number of small objects: the asteroids.

The primitive earth experienced a tremendous churning of matter and energy: meteorites, volcanic eruptions, rains, storms, ultraviolet radiation, radioactivity. Volcanoes spewed enormous quantities of gases, which formed the primitive atmosphere. Water vapor was present, but no oxygen.

It was under these extreme conditions that life must have taken its first steps.

miles) from the sun and not at 150 million kilometers (93 million miles)—it would have followed the fate of Venus. It would have choked in its own atmosphere; water would have evaporated, leaving a desiccated and burning soil. On the other hand, if the earth had been farther away from the sun—for example, 160 million kilometers (100 million miles)—the soil would have cooled, and numerous ice caps would have gradually covered most of the globe.
If the earth had been a little farther or a little closer, there would have been no blue planet!

J.-L. Alart

The Components of Life

Chemical analysis of the matter in living beings shows that it consists of molecules. A molecule is a combination of atoms. We talked about atoms earlier, describing how they were formed and how they created the earth. Everything alive—from a bacterium to an elephant, wild flowers or fish—consists of molecules of a rather specific type. They are called organic molecules, which means that they contain carbon atoms. But a simple mixture of organic molecules is not life: it lacks organization.

Organic molecules such as cellulose, fats, and proteins may consist of hundreds or thousands of atoms. Some of these molecules play a fundamental role because they organize living matter. One of the most important groups of such molecules are the proteins. Proteins consist of long chains of amino acids. About twenty different amino acids exist. But they all have a common characteristic: the same coupling system. Similarly, deoxyribonucleic acid—DNA—and ribonucleic acid—RNA—are very long chains consisting of only four kinds of links: the four nucleic bases. Nucleic bases and acids are so named because they are found in the nucleus of the cell.

This system resembles a game of construction blocks. The twenty amino acids and the four nucleic bases are called the "building blocks" of living matter.

The Primitive Soup

Living beings actually make the molecules of which they are composed themselves. The first question, then, concerning the origin of life is: How were the first chains of mole-

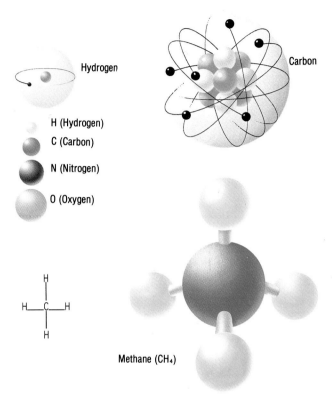

Hydrogen

Carbon

H (Hydrogen)

C (Carbon)

N (Nitrogen)

O (Oxygen)

Methane (CH₄)

An atom consists of a nucleus (protons + neutrons) surrounded by elec-
trons. Molecules are combinations of atoms. The simplest organic
molecule, methane, is formed by four hydrogen atoms bonded to a carbon
atom. Methane is commonly used in camping stoves.

cules made before the existence of living beings?

Because one finds traces
of life in rocks almost four
billion years old, it seems
that the first building
blocks of living matter must
have appeared in the en-
vironment of the primitive
earth.

Around 1950, in the
University of Chicago
laboratory of Harold Urey
investigators were trying to

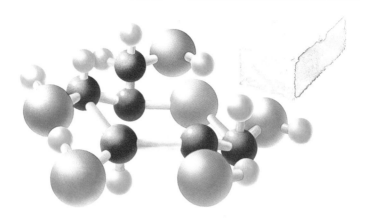

A rather simple organic molecule: glucose. *It consists of six atoms of carbon, twelve atoms of hydrogen, and six atoms of oxygen ($C_6H_{12}O_6$). (Sugar cane is made of glucose molecules.)*

determine the composition of the primitive atmosphere. They thought it should have consisted mainly of methane, hydrogen, ammonia, and water vapor. In 1953, Stanley Miller, one of the investigators, decided to attempt an experiment. He built a very simple apparatus to try to reconstruct in a few days what had taken place during millions of years on the primitive earth. In the experiment an artificial atmosphere of methane, hydrogen, ammonia vapor, and water vapor was subjected to electrical sparks. Miller then analyzed the results of his experiment, and his findings were as-

tonishing. Numerous molecules had been formed, among them amino acids, the building blocks of organic matter. Miller then theorized that amino acids could have been formed by lightning discharges in the primitive atmosphere.

Proteins, a game of construction blocks:
1. *Each building block, an amino acid, is an organic molecule. There are twenty different blocks having the same coupling system. When two blocks are in the right position, they couple. Thus very long chains of amino acids are formed: the proteins.*

Nucleic acids:
2. *There are four different nucleic bases. Like the amino acids, they too have a coupling system. Thus extremely long chains are formed: DNA and RNA.*

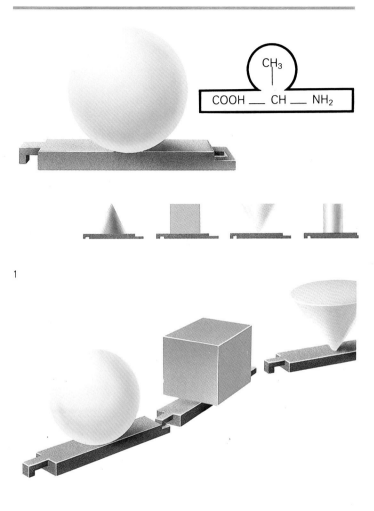

1

2

MILLER'S EXPERIMENT

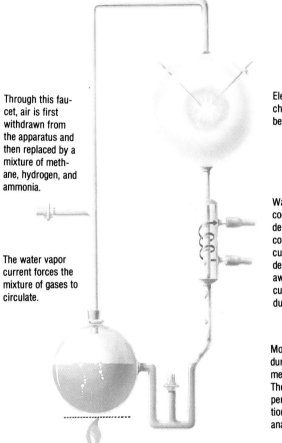

Through this faucet, air is first withdrawn from the apparatus and then replaced by a mixture of methane, hydrogen, and ammonia.

The water vapor current forces the mixture of gases to circulate.

PRIMITIVE ATMOSPHERE

CROSSED BY LIGHTNING
Electric spark discharge takes place between electrodes.

RAIN
Water vapor is cooled by condenser in which cold water circulates. It condenses and carries away the molecules being produced.

OCEAN
Molecules formed during the experiment collect here. The stopcock valve permits the collection of samples for analysis.

This balloon vessel contains water that boils continuously during the length of the experiment.

Stanley L. Miller completed his experiment *using very simple equipment that can be found in any chemistry laboratory: glass vessels, tubes, and stopcock valves. He kept his experiment going continuously for eight days. Water boiled and electric sparks discharged between the electrodes inside the glass balloon vessel filled with simple gases. The experiment was constructed to simulate the particular conditions prevailing on the primitive earth.*
In the lower tube, the liquid became red and cloudy: the simplest part was over. It was now necessary to analyze the products obtained, a long and delicate work, but one that yielded very surprising results. Numerous organic molecules were found, including amino acids.

A great step forward had been made, even though in Miller's experiment only two or three amino acids were obtained among many useless molecules. Great excitement spread throughout the scientific world. People wanted to know what would happen if the source of energy or the composition of initial gases were changed. During the sixties, John Oro in Texas,

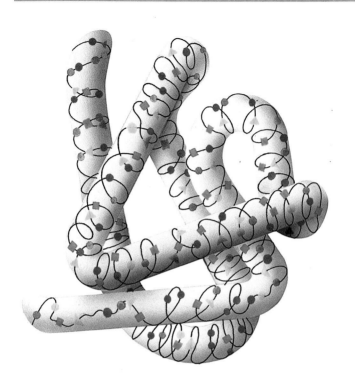

In a game of construction blocks, all kinds of shapes can be obtained from different pieces by changing the assembly order.
A protein is a chainlike combination of amino acids. Depending on the order of the amino acids, very different proteins are made, each one with a precise function.

Melvin Calvin in California, and many others succeeded in obtaining not only the twenty amino acids and the four nucleic bases, but also sugars and fats. All the molecules formed in the primitive atmosphere must have been carried by rain into the sea. This warm sea, which can be visualized as having been full of organic molecules, is called the "primitive soup."

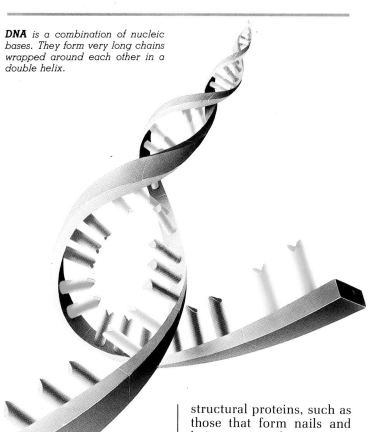

DNA is a combination of nucleic bases. They form very long chains wrapped around each other in a double helix.

From Block Molecules to Chain Molecules

Proteins are the main constituents of living beings; they can perform almost any function. There are structural proteins, such as those that form nails and hair; contractile proteins, such as those found in muscles, and—above all—enzymatic proteins (also called enzymes), which control all chemical reactions, such as, for example, those involved in digestion.

A protein is a chain of amino acids. Individual amino acids were isolated from one another in the simulated primitive soup. To come closer to what occurs within a living being, the formation of chains of molecules is indispensable. Similarly, nucleic bases must develop chains to form DNA or RNA.

Several Possible Scenarios

The formation of chains is difficult in the presence of water. Several scenarios were proposed for the primitive soup to become concentrated and to dry out. The soup may have evaporated under the sun along a shoreline or in contact with hot volcanic lava; or it may have deposited organic molecules on layers of clay, thus promoting the formation of chains. Experiments have shown that any of these scenarios could have occurred.

For instance, Sidney Fox obtained amino acid chains artificially simply by heating the amino acids. He called the chains he produced proteinoids to avoid confusion with proteins found in living beings.

In summary, we can conclude that the components of life appeared before life. Thus a small part of the

problem of life's origin has been solved. However, these life components in the primitive soup were still in total disorder: there was only a soup.

NATURAL HISTORY MUSEUM SHOP

33 CASH-1   0893 0002 203

CLIMATES   MDS 1N   4.95
VOLCANOES   MDS 1N   4.95
ORIGINS   MDS 1N   4.95
20 YEARS   MDS 1N   4.95
ENGINEERN   MDS 1N   6.98
                   TOTAL   26.78

          CASH TENDER   27.05
          CHANGE   .27

THANK YOU PLEASE VISIT OUR OTHER SHOPS

8/07/88   14:35

The Orion nebula. *The light-colored areas consist of gases emitting light. The dark-colored areas are clouds of dust particles masking this light. In order to observe interstellar molecules, it is necessary to collect the radiation they emit. It is rarely visible light. Most commonly, the radiation consists of short radio waves that can be detected by large antennae. Waves are emitted when molecules impact with one another and are characteristic for each molecule. It can be said that each molecule has its own signature—radio wave characteristic—that allows its identification.*

31

The American scientist Sidney Fox assumed that a small puddle of primitive soup at the foot of a volcano may have started to boil and then dried out. Therefore, he heated a mixture of amino acids in an oven at 170° C (338° F) for several hours. The result was not a horrible black residue. On the contrary, chains of amino acids, as in a protein, had formed.

Radiotelescope. Large radiotelescopes such as this are used to investigate the presence and distribution of molecules in interstellar space.

Organic Molecules in Space

*Less than fifty years ago, some organic molecules
were discoverd in space by means of optical
telescopes. Subsequently the systematic collection of
radio waves emitted in space permitted the identification of many of these molecules. The small collection
of molecules increased. Many of these molecules —
often very simple, although some consisted of thirty
atoms — were detected in clouds of gases and dust
particles and were identified as similar to the ones
that gave birth to our solar system. Numerous radiotelescopes particularly suited for this research are
now operational throughout the world and more dis-
coveries are expected.*

*The question was raised as to whether these mole-
cules could have shaped the environment in which
the building blocks of life had formed. Some of the
molecules in interstellar space were complex, but
no amino acids were found. However, could interstellar molecules have reached the earth to serve as
elements for the first molecules of living matter? Probably no. The condensation of interstellar clouds to
form the sun and the planets was so violent that the
molecules could not have survived the very high
temperatures existing at the origin of the earth.*

Toward Life

The study of an extremely simple living being—for instance, a bacterium several thousandths of a milli- meter in size—shows that life is essentially self-orga- nization. Billions of different molecules react with one another under the control of hundreds of pro- teins. Even the production of proteins is controlled and regulated by proteins. Everything controls and regulates everything.

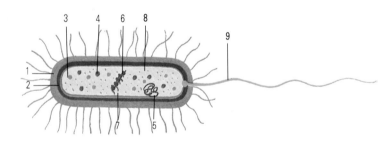

A bacterium. *This organism appears to be one of the simplest forms of life, but biochemists have discovered that a bacterium is, in fact, extremely complex. 1. Cell wall (formed by chains of sugars). 2. Cell membrane (dou- ble layer of fatty substances). 3. Proteins. 4. Ribosome (system of transcrip- tion). 5. Nucleic message (DNA). 6. Fragment of nucleic message undergo- ing transcription (RNA). 7. Protein being formed. 8. Protoplasm (billions of molecules involved in billions of chemical reactions). 9. Flagellum.*

Proteins are not built in a random fashion. Each amino acid has its own characteristics. If amino acids are linked in a certain order, the resulting protein will have a given shape and will play a given role. Proteinoids, formed according to the technique of Sidney Fox, are also chains of amino acids, but without any particular order. Their shape is random and they have no precise function. In living beings, each protein is a small machine that fulfills a single specific function. An enzyme, for example, has a well-defined shape and controls a specific chem-

Each enzymatic protein is a machine to catch certain molecules. It recognizes them accurately, attaches to them in a lock-and-key fashion, and then releases them.

In order to take in carbon, which exists in the air as carbon dioxide (CO_2), a plant has a system of enzymatic proteins that surrounds chlorophyll. The proteins bind carbon (C) to an organic molecule, producing a carbohydrate, and release oxygen (O_2). This is the way—the process of photosynthesis—that a plant constructs organic matter and grows.

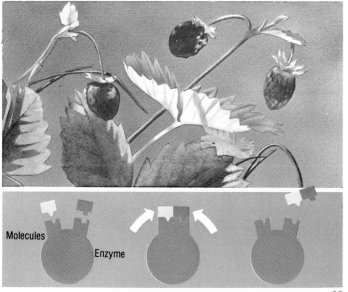

Molecules

Enzyme

A living being has a stable organization. *Nevertheless, inside the organism everything changes and moves constantly. This can be compared to a whirlpool of a river, which maintains its shape while the water forming it changes continuously.*

ical reaction that breaks and reassembles certain molecules.

A Membrane to Protect the Internal Environment

The first requirment of a living organism is to find means of protection. It is the minimum requirement for organization. An or-

ganism needs an envelope, a membrane. Inside, well-organized chemical reactions provide a certain independence with respect to the outside. Nevertheless, the membrane should not isolate the living being completely. A *closed system* completely isolated from the outside, wastes away and falls into complete disorder. A living being, on the other hand, is an *open system*. It requires matter and energy for survival and growth. The membrane of a cell makes choices: it allows some molecules to enter and waste products to exit. Inside, it retains proteins.

Like Soap Bubbles

In 1965, Sidney Fox, whose experiments produced proteinoids, made a new and very interesting discovery. When he placed several milligrams of proteinoids in water, millions of *microspheres* were generated. A microsphere is a type of bubble full of water. Everybody is familiar with soap bubbles. In reality, they are air bubbles with an envelope of soapy water floating in air. In microspheres, proteinoids form the wall of the bubble. This wall has a double-layered structure that closely resembles cell membranes. It is semipermeable, letting certain small molecules enter and keeping larger ones out. Therefore, a microsphere, like a cell, is an open system.

Microspheres. *Under the electron microscope, one can easily distinguish the double layers of proteinoids that form the envelope of these minute bubbles, which resemble soap bubbles.*

Microsphere

Double layer
of proteinoids

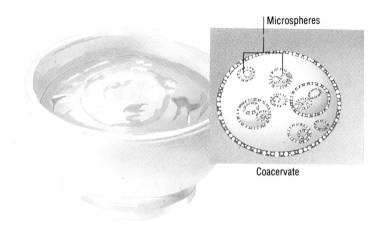

Microspheres

Coacervate

Everybody knows how to make vinaigrette or mayonnaise. When stirring a mixture of oil and vinegar with a fork, the vinegar separates into droplets which are immediately covered by oil. If one adds mustard and salt while stirring, one causes still smaller droplets of the different ingredients to enter the vinegar. This is comparable to microspheres, which become enriched in various molecules and increase in complexity. These encasings of microspheres are called coacervates.

The primitive soup contained not only proteinoids but also sugars, fats, nucleic bases, etc. As they became increasingly embedded in the bubbles of proteinoids, these molecules formed small pockets that rendered the structure more complex. Microspheres become encased one within the other. Something similar happens when one beats vinaigrette or whips mayonnaise vigorously.

An Exciting Scenario: The Appearance of Life

As early as 1924, the Soviet scientist Oparin, who had studied the structure of microspheres, noticed that in a system of complex microspheres millions of different organic molecules are clustered together in what is probably a place of intense chemical activity.

These microspheres grow and eventually break up.

They form new micro-spheres in which molecules are scattered at random. Those containing the best combinations are pre-served and grow. A kind of selection takes place in which only the most resis-tant microspheres "sur-vive."

But is this life? Is it even a path toward life?

According to certain scientists, this scenario is a good one. In fact, the most important steps have been taken. An organization ex-ists that is capable of self-preservation. In their opinion, such microspheres will improve. They will eventually acquire the same system of reproduction that present organisms have.

Coccoids and spheroids *extracted from a rock one billion years old. On the left, an electron-microscope picture shows several coccoids enlarged 45,000 times. What looks like a small mouth must have been an opening used for reproduction. On the right, a light-microscope picture, enlarged 25 times, shows a large spheroid consisting of thousands of coccoids.*

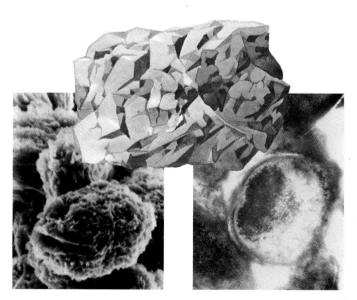

Looking at the Oldest Fossils

From Coccoid to Bacteria

In rocks older than three billion years, paleontologists discovered small spherical forms resembling microspheres, although slightly larger: the "coccoids."

These coccoids resemble some living bacteria. If indeed coccoids evolved into bacteria, a stage is missing that remains very mysterious: the stage that led coccoids to reproduce in the same manner as living bacteria.

From Spheroid to Our Cells

In these same old rocks not only do isolated coccoids occur but also coccoids associated in chains, in masses, and even in masses surrounded by a membrane. These very delicately organized groups of coccoids are called spheroids.

Here things appear somewhat clearer. One can often observe spheroids ejecting a kind of substance through a small opening. It is probably a jelly full of coccoids. From this jelly new spheroids would form. This process recalls a well-organized reproduction system.

Life: Russian Dolls?

What have we found since the first organic molecules? Step by step the combina-

Method for extracting microfossils. *From certain rocks one can extract fossils so minute that they are invisible to the naked eye. The rock, after having been crushed into a very fine powder, is mixed with water. When one blows air into this liquid with a pipette, a froth forms at the surface. In this froth, microfossils are concentrated. They can then be observed under the microscope.*

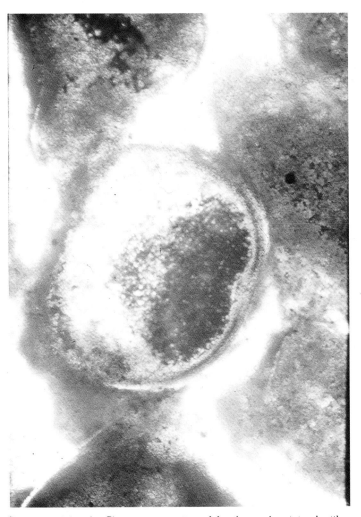

Staining of fossils. *The organic matter of fossils can be stained with a special chemical. Wherever a red color occurs, one can assume the existence of organic matter. Therefore, the little round dots are not simply pebbles; they are microfossils, namely thousands of small coccoids. They can be seen clustered at a pole that will open and serve for reproduction.*

Simple molecule (atoms bonded together)

Amino acid (complex molecule)

Proteinoid (combination of amino acids)

From atom to living being, at each stage organization increases.

tion of simple elements has provided new qualities to the whole. We come to think that perhaps life emerged through a nested system.

—Atoms form molecules such as amino acids.

—Amino acids form chains and generate proteinoids.

—Protenoids gather to form microspheres, then coccoids.

—Coccoids combine to form spheroids.

The spheroids would be the origin of our cells. A body like ours consists of thousands of billions of cells.

At each stage, the whole is more than the sum of its parts. For instance, a proteinoid is more than the sum of its amino acids because it can have a precise function within a microsphere. A microsphere is more than the sum of its proteinoids because it can have encompassing reactions signifi-

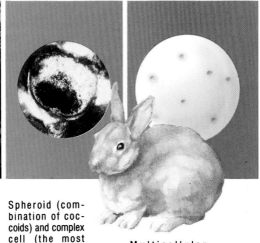

Microsphere (combination of proteinoids) and coccoid (simplest fossil form)

Spheroid (combination of coccoids) and complex cell (the most general living form)

Multicellular organism (combination of complex cells)

cant for its own survival.

But is all this so extraordinary? Everyone knows, for example, that a discussion among several persons produces more interesting ideas than those reached by a single individual.

Divided Opinions

We have seen that life might be the final product of an increasingly developed self-organization. The reproductive system would be only an aspect of this self-organization and would have developed gradually during evolution.

Many scientists disagree completely with this way of thinking. According to them, the problem of reproduction is not trivial. On the contrary, they think that self-reproduction is the most important characteristic and that it should exist before any definite organization. Any living form must be capable of reproducing before it can improve and organize itself.

The Enigma of Self-Reproduction

Living beings are organized according to a precise blueprint. Indeed, each organism has within it a message that indicates precisely how it is constructed: the *genetic code*. The blueprint of the organism is recorded in a chain of nucleic bases: the DNA located in the nucleus of the cell.

DNA is like a huge book written in very small characters on a single and extremely long line. The genetic language has only four different letters: the four nucleic bases.

As in our language, the order of these letters is important. For instance, the following words have the same letters but in a different order:

A MESSAGE

MAG ESSEA

The meaning of the message depends on the order of the letters. Similarly, the meaning of the genetic code depends on the order of the nucleic bases.

Duplication

At the time of reproduction, the blueprint of the

Single-stranded DNA chain. *Along the DNA molecule, nucleic bases are linked together. The four kinds of nucleic bases are the four "letters" that allow the writing of the genetic code.*

parents (DNA) has to be transmitted to their offspring. This is why children resemble their parents. When a bacterium reproduces by division, the two daughter bacteria are identical to the mother (which they replace). The genetic code of the mother has been "duplicated"— that is, copied and transmitted. Thus the code is preserved and multiplied.

The astuteness of the DNA molecule lies in its system of correspondence between nucleic bases. This allows the code to reproduce itself, much as a book might print itself.

Transcription of the Message

Since the genetic code expresses the blueprint of the organism, it must be read, understood, and applied. This is *transcription.*

Double-stranded DNA chain. *Different nucleic bases correspond to each other—two by two, face to face. DNA occurs as an extremely long double chain, which may reach more than a meter in length in higher animals.*

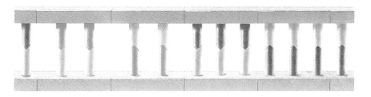

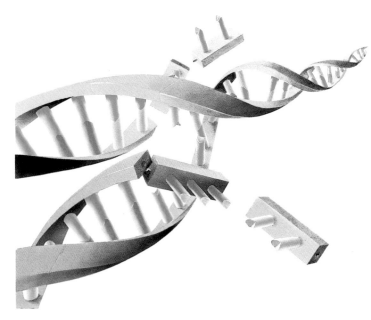

Duplication. *The DNA molecules consists of two strands. When the strands separate, new bases position themselves opposite corresponding bases. Thus, two new double chains—identical to the first one—are formed. The figure shows the ongoing process.*

During transcription, the proteins of the daughter bacteria are generated by following the blueprint inherited from the mother. The *genetic code* allows this operation by means of a very clever process.

Transcription proceeds gradually along the DNA. Each group of three nucleic bases codes for a certain amino acid. As these specific amino acids "line up" and connect to one another in the specific order directed by DNA, a chain of amino acids—a protein—is gradually formed. Each portion of the DNA molecule corresponds to a different protein. This is how the blueprint of all proteins forming a living being is transcribed in the genetic code.

We have seen that each protein with its precisely ordered amino acids is a small machine that performs a definite task. The extraordinary feature is that the very actions necessary for transcription are performed by proteins specialized in this particular task. Are these proteins recorded in the genetic code? One can imagine the following circular dialogue:

—"Transcription proteins have to be transcribed.

—By what?

—By other transcription proteins.

—Where do they come from?

—From the transcription of the genetic code.

—By what?

—By other transcription proteins."

There is no way out unless at the beginning, a genetic code and proteins capable of transcription were present. This situa-

Genetic code. *This is the rule that determines which amino acid corresponds to a given sequence of three nucleic bases (triplet). The same triplet of nucleic bases can never correspond to two different amino acids.*

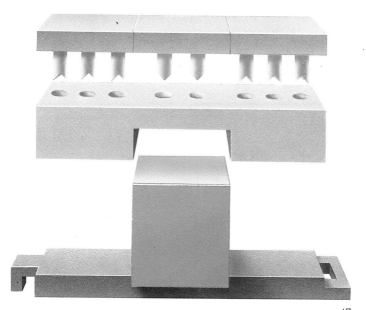

47

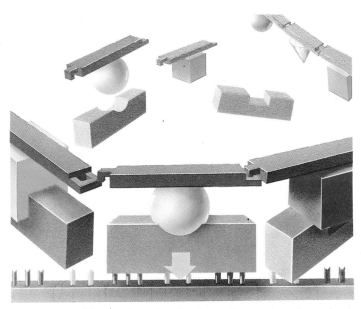

Transcription. *Along a chain of nucleic bases, the genetic code constructs triplet by triplet a chain of amino acids, namely a protein.*

tion can be compared to the following: we would have a book that would tell how to construct a robot capable of reading this book in order to construct such a robot. Obviously a first robot is required to read this book!

A Beginning . . . and Evolution

Let us imagine that the formation of a small portion of DNA and of some tran-

scription proteins can be explained. Would we have the key to the mystery? Yes, because everything might improve, and history could go like this: A small change—a mutation—is produced randomly in the DNA. For instance, a nucleic base is replaced by another. The new code is a little different. The new protein, formed during transcription, is somewhat modified. If the resulting organism does not function as well, if it is more fragile, it will disappear. But if the change represents an advantage—if, for instance, duplication is more rapid with the new protein—then this new code will multiply and be preserved in the next generation. This is evolution. The blueprint and its system of transcription would have gradually improved to become eventually those of a modern organism.

The problem of life is solved, therefore, if one has a beginning of *duplication* and a beginning of *transcription*.

A Beginning of Duplication?

Is it possible to obtain a chain of nucleic bases without the help of life?

When a hand draws another hand, it produces it. *But for a hand to draw the next one, it must already exist. How did it all start? The origin of reproduction of living beings resembles the paradoxical system of two hands sketching each other at the same time. This question is still alive.*

Some scientists have shown that it is feasible. In contact with clays, nucleic bases concentrate and link together in interstitial spaces. Furthermore, the natural tendency of nucleic bases to combine in pairs should cause the formation of double chains. In order to obtain true duplication, a few obstacles still exist, but answers are possible. However, one runs into the same problem encountered with proteins. Nucleic chains produced in the laboratory have no par-

Mineral "Life" Before Organic Life?

We may admire the complexity of a cathedral. However, it is inconceivable that it was constructed by putting side by side the arches and vaults over an empty space. We know that it was built from a wooden scaffold constructed by simple means, which have nothing in common with the assemblage of stones. Is it possible to imagine that something similar happened with living things? Are there inorganic structures in nature that can carry a complex code (a blueprint), that can multiply without the help of other structures and serve as a scaffold for chains of organic molecules? Cairns-Smith, a chemist at the University of Glasgow, believes that clays may have played such a role. They are present everywhere on the surface of the earth in the form of microscopic crystals.

Crystals consist of atoms usually arranged in perfectly regular networks. In reality, they often contain many defects (atoms which are not exactly in their assigned position). Once a defect appears, it is repeated during the crystal's growth. If the crystal breaks up under the action of some exterior force (for instance, some

ticular order. They have no meaning. They are like books with randomly printed letters. Proteins obtained from these nucleic acids by transcription would have no defined activity.

A Beginning of Transcription?

It was believed for some time that the problem was relatively simple. If each group of three nucleic bases had a natural tendency to attract a specific

clays separate like sheets in a stack of paper), small crystals preserve the pattern of defects and grow according to it. One might say that reproduction and preservation take place—something that resembles life.

Over a period of time, a structure of chains of organic molecules might have developed over a complex scaffold of clays that preserved their pattern of defects. This is comparable to the cathedral's stones resting on the wooden scaffold. During evolution, chains of molecules could nave acquired more well-defined and complex forms and might no longer have needed a scaffold in order to duplicate.

amino acid, then amino acids would naturally congregate along the DNA chain and would form the corresponding proteins. Unfortunately, all experiments to prove such a tendency have failed.

Is the system of transcription then purely random? A series of extraordinary coincidences would be necessary. First, a perfectly ordered nucleic chain. Second, chains of amino acids forming by chance proteins capable of transcription. Third, the encounter of these two types of chains of molecules!

Faced with the immensity of the problem, some biologists believe that the mystery is beyond understanding and that life is a miracle.

But this is not considered reasonable by some people doing research. Many scientists, therefore, try to find some hypothesis to open the door to new experiments. They think that the complexity of the two types of chains (nucleic message and proteins) can be explained if one assumes that they evolved together. The evolution of one would help the evolution of the other and vice versa: a coevolution, similar to progress in technology. The first cars needed, for instance, improved roads; these roads allowed in turn the construction of better and faster cars, etc.

A Clay Mold

Another very promising field of research lies in clays. We mentioned above that clay crystals promote simultaneously the formation of chains of amino acids and nucleic bases. Crystals in successive sheets might have served as "molds" for the two types of chains, perhaps even establishing between them a correspondence leading to the origin of transcription.

The above, not so fruitful attempts to explain the origin of self-reproduction show that the problem persists. Certain aspects of the beginning of life remain an enigma.

We should perhaps change entirely our way of reasoning in regard to all previous paths of investigation. We must find a new logic, a different way of looking at seemingly paradoxical things, and take a completely fresh approach to life.

Clays in Support of Life

At the surface of the earth, in contact with the atmosphere, rocks undergo weathering. Rainwater dissolves constituents, transports them, deposits them farther on, and evaporates. Microscopic crystals appear: these are the clays.

Clay crystals are similar to stacks of thousands of sheets. Each sheet consists of various atoms arranged in a regular and distinct manner. Between each sheet is a space. This space, however, is sufficient for foreign molecules to slip in. It is possible that this space played the role of a mold for the production of long chains of organic molecules.

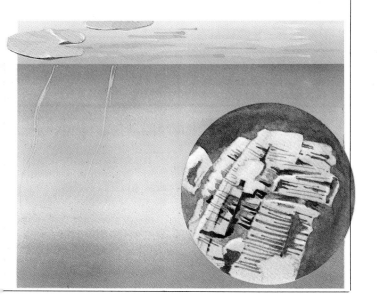

Does Life Exist Elsewhere?

Since antiquity, numerous naturalists and philosophers have believed that other inhabited worlds exist. They believed that what surrounds us must exist elsewhere, and perhaps even in infinite number. Today, with the progress of ideas about the beginning of life on earth, improved means of observing the universe, and greater knowledge about planets and their formation, the study of extraterrestrial life is no longer a matter of belief; it is a question that can be approached in a scientific manner.

In the Solar System?

Probes sent to Venus by the Soviet Union and the United States have shown that this planet is completely inhospitable to life, with a ground temperature close to 500°C (932°F).

But Mars raised much hope. The planet Mars resembles the earth. In the nineteenth century, Schiaparelli, an Italian scientist, and Lowell, an American, believed they saw gigantic canals crisscrossing the surface of Mars. This meant that living beings with a technology more advanced than ours were active! It did not take long to understand that the supposed canals were an optical illusion. Nevertheless, Mars deserves closer study. Probes sent by NASA contained instruments capable of finding possible traces of life. Nothing conclusive was found, and the experiments were much disputed.

A curious object exists in the solar system. This is Titan, the largest satellite of Saturn. It resembles a primitive and very cold

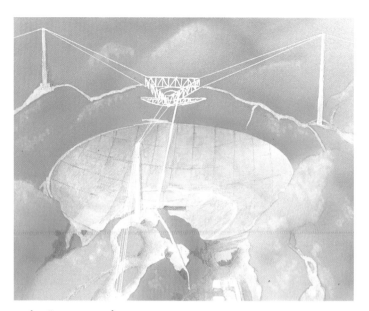

earth. Its atmosphere undoubtedly contains organic molecules, such as those that preceded life on earth. Titan is so interesting that a visit is scheduled for the nineties.

In Our Galaxy and Beyond?

If no living beings are encountered in the solar system except on earth, it would be necessary to search farther away, beyond the sun, toward other stars. This is being done by scientists interested in ex-

The radiotelescope at Arecibo (Puerto Rico) is built in a natural depression with a diameter of 307 meters (1000 feet). A message addressed to possible extraterrestrial beings was sent from this instrument in 1974. This radiotelescope also receives possible messages.

traterrestrial life. They are attempting to evaluate how many planets in our galaxy can sustain life—a very difficult evaluation indeed! Billions of stars that resemble the sun exist, but we have only very vague ideas about the existence of other planetary systems. Even if such systems exist, how many planets that resemble earth can be found in them? Evaluation of the number of "earths" capable of sustaining life varies from one (ours) to a very large number. One hopes that in the not too distant future very sophisticated telescopes, capable of detecting planets around some of the closest stars, may be sent into space.

Listening to the Sky

Meanwhile, some scientists assume that life exists elsewhere and that it has evolved and produced intelligent beings who have the means and the desire to communicate. Since 1960, tens of observers have used radioastronomy antennae to listen to the sky for many hours, days, and even months. Until now, nothing has come of it. But new methods exist, as well as spectacular plans for listening to possible extraterrestrials. Still other scientists have sent messages, hoping that they will be received by extraterrestrials (if they exist) and that the latter will understand that these messages come from afar and deserve to be studied

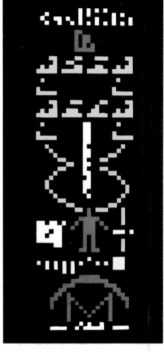

Message sent by the antenna at Arecibo, November 16, 1974. *It contains various information about the solar system, the population of the earth, DNA molecules, and the main chemical elements of living matter.*

A: Symbols for hydrogen, oxygen, nitrogen, carbon, and phosphorus.
B: DNA code.
C: DNA helix.
D: World population; man and his size.
E: Solar system.
F: Antenna at Arecibo.

with care. But let us not forget a small detail. In order to reach the closest star, a radio message takes more than four years. And most stars are much farther away. If there should be an answer, let us be patient!

The Domain of the Imaginary

Our hope of communicating with extraterrestrials means that we imagine them to be very similar to ourselves. Yet, even on earth, there are writings of vanished civilizations that have not been completely deciphered! If life elsewhere evolved differently from life on earth, it is perhaps futile to try to communicate with it.

Moreover, some scientists think that life elsewhere might exist under lesser constraints than those mentioned in the first chapter. Thus one can imagine nonchemical forms of life—for instance, organisms with bonds different from those that link atoms together. But here we enter an imaginary domain, although scientifically conceivable.

Space probes collected fungi spores at high altitudes (38 miles, or 60 kilometers). Desiccated, these spores may resist solar radiation. Is the solar wind able to blow them away on an interplanetary trip?

Panspermia, or Seeding of the Universe

Experiments by Pasteur led to the belief that living organisms could have been born only from other living organisms. At the end of the nineteenth century, several scientists explained the beginning of life on earth, as originating from germs that came from outer space. This was the theory of "panspermia." Living worlds were assumed to seed each other.

The French scientist Becquerel disagreed. Living germs could not resist interplanetary voyages because they would be killed by radiation, which is everywhere in outer space.

More or less abandoned, the theory of panspermia was again promoted vigorously fifteen years ago by the famous English astrophysicist Fred Hoyle.

Here is what Hoyle said: Chemical processes of living things are extremely complex. They could not have become established in some hundred million years. Much more time was necessary as well as a place less narrow and less isolated than our mere terrestrial globe. Resources from the entire galaxy were needed, from all its hundred billion stars, in order to produce the simplest cell. All interstellar clouds, everything that resembles a planet from far away or at close distance, had to be involved in the beginning of life. Comets are the seeding agents of the universe. They deposit, in the vicinity of planets, organisms that are more or less complex, and perhaps even bacteria that resist ultraviolet radiation or X-rays.

The evolution of living beings on earth is assumed to result from the addition of the genetic code of new life forms coming from outer space....

This interesting theory gives a much more universal meaning of life. It suggests that we should no longer believe that we are the center of the universe in a biological sense, just as Copernicus and Galileo taught us that we are not the center of planetary movements.

The theory of the seeding of the universe by living beings cannot be rejected for at least one reason. We have no proof that life is not as old as the earth, or even older. We know only that life on earth is at least 3.8 billion years old.

Has man started to spread life in the universe? *In spite of all precautionary measures, have probes sent to the moon, Mars, Venus, and to the far reaches of the solar system carried germs of terrestrial life?*

Life at a Depth of 10,000 Feet

Geologists are responsible for an astonishing discovery. They brought back from the bottom of the ocean (from a depth of 7,000 to 10,000 feet, or 2000–3000 meters) photographs of colonies of entirely unexpected animals—fish, crustaceans, worms, sea anemones—that live in complete darkness. These colonies live very close to "hydrothermal springs" of very hot water (250°C, or 482°F). Bacteria live in this water. Their presence indicates that life can exist in an environment that is extremely different from the one we know at the surface of land or sea. Some biologists emphasize that hydrothermal springs combine all the necessary conditions for development, from the simplest molecules to living organisms. Water flowing out of these vents is seawater that infiltrated through fissures. It came very close to the magma, at 1000°C (1832°F). It dissolved many chemical components and then rose through other fissures, cooling gradually. One can thus imagine the following scenario. At the greatest depth, simple molecules are formed: methane, carbon dioxide, etc. Higher up, organic molecules, the building blocks of living things are constructed. Still higher up, on clays that cover the walls of fissures, molecular chains are formed. Finally, almost at the outlet of the springs, pro-toorganisms appear. Thus, life might begin around a "hydrothermal chimney". Hydrothermal springs were very numerous at the beginning of the earth; could they represent the privileged place for the origin of life? And since these springs still exist, does life continue to begin there?

Explosion of Life

Let us look around us at the living world, at plants and flowers, grass and wheat in the fields, trees, herbivorous and carnivorous animals, earthworms, fish in the seas, and birds in the sky.... What diversity! What swarming of living forms! Among insects alone, there are more than one million different species!

During millions of years, living beings never stopped changing and evolving. After coccoids and spheroids came primitive fungi, sponges, strange squids, and the famous trilobites. Bony fish established themselves in the seas 500 million years ago; then, creeping plants settled on continents. Later on, forests of giant ferns spread, which were invaded by the first reptiles. Then came the reign of the dinosaurs and flying reptiles, during which the first mammals appeared. Much later, close to the age of the mammoths, the first traces of man were found.

The study of fossils is thus very important in understanding the history of life. Many species have disappeared, while others have appeared. Classifica-

Chimera. *Man has always loved to imagine a combination of living forms, imaginary creatures such as the sphinx or the flying horse Pegasus. These combinations are called "chimeras." Spheroids are real biological chimeras, resulting from a combination of coccoids with different capacities.*

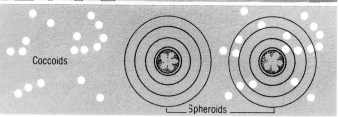

Coccoids

Spheroids

tion of these fossils according to their order of appearance shows that various species of a certain epoch evolved from an earlier species, which, in turn, originated from an earlier one.

Thousands or Billions of Specialized Cells

Evolution allows us to recognize the most important divisions of the living world. A fundamental division separates unicellular organisms from multicellular ones. Microbes such as bacteria consist of one autonomous cell. Plants and animals, however, are formed from thousands or billions of cells. Furthermore, each of these cells is much more complex than a bacterial cell.

Each cell of a multicellular organism has a nucleus, which contains the genetic message; mitochondria, which supply the cell's energy needs; and other more or less complex structures.

Each one of these complex cells can specialize. For instance, in a multicellular organism like ours, there are skin cells, muscle cells, nerve cells, and others. In a plant, root cells

absorb water from the soil, cells of the stem have hard walls, and cells of leaves contain chlorophyll, which captures solar energy.

Is the Origin of Complex Cells a "Chimera"?

We can trace the origin of organisms to simple cells such as bacteria. But the first important step in evolution was the appearance of the first complex cell

More and more, this is thought to be a "chimera." In antiquity, imaginary animals, formed by combining parts from several animals, were called "chimeras." Examples of these are the centaur—half man, half horse—and the sphinx—a fabulous creature with a human head and the body of a lion.

The "chimera" that supposedly led to the first complex cell could have been an association of simple parts. Reorganization of these parts might have happened in the following manner. Coccoids evolved into very simple cells such as bacteria. Living under different conditions, they differentiated and formed various species. In the

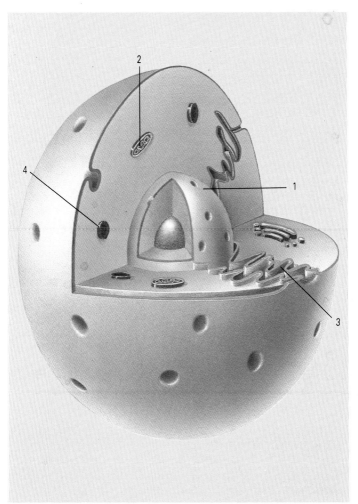

The eucaryotic cell. *What we presently call a "cell" in a complex organism is, in fact, the "eucaryotic" cell, from the Greek "cell with a true nucleus." One can see that this cell is very different from simple and isolated cells such as bacteria (see p.* *(see p.   )**). In addition to a nucleus (1), it has mitochondria (2), an endoplasmic reticulum (3), lysosomes (4), and other parts not illustrated here.*

center and around a spheroid, various species of coccoids concentrated. This concentration allowed a combination of their capabilities.

A complex cell has qualities superior to those of its individual components. It has above all the advantage of being able, when multiplying, to form even more complex and better adapted multicellular beings.

Concrete evidence supports all of these hypotheses. First, cells of some fungi, more primitive than those of flowering plants, have many characteristics in common with those of spheroids found in rocks a billion years old. Second, present complex cells preserve traces of their history. For example, mitochondria, cytoplasmic components responsible for the respiratory functions of the cell, are probably former bacteria. They contain indications of an earlier autonomy in the form of small pieces of genetic message.

What Is the Difference Between Plants and Animals?

The second great division of the living world separates plants and animals.

The very first organisms took their food directly from organic molecules formed in the sea where they lived. However, this food must have rapidly run out. It was therefore necessary for living organisms to produce their needed organic matter directly. This is what existing plants do. They possess a system of chlorophyll and enzymatic proteins that allows them to use solar energy to extract carbon from carbon dioxide in the air (see p. 37) and make sugars, amino acids, and nucleic bases. Herbivorous animals, eating the plants, consume ready-made organic molecules. Finally, they themselves are eaten by carnivorous animals...or they end up on our plates.

Life is supposed to have started as a very simple animal form that ate organic molecules contained in the primitive soup. These first organisms then evolved gradually to produce the first plant forms: very simple unicellular algae.

Plants produce their own organic molecules by using solar energy. The rabbit eats plants to make its own organic matter. A carnivorous animal such as the fox eats the rabbit and assimilates this organic matter. Humans, who eat everything, will eat either a simple vegetable soup, a rabbit, or a carnivorous animal.

Earth Changes Its Atmosphere

The appearance of the first algae was of extreme importance for at least two reasons. First, these chlorophyll-bearing cells were an important source of organic matter. Organisms capable of feeding on this organic matter appeared soon thereafter. These were the first true animals.

Second, production of organic matter by plants released oxygen. Let us recall that there was no oxygen in the primitive atmosphere. The first plants thus gradually formed the air we breathe today. What a change in the conditions of life on earth! There was not only air for animals to breathe but also a layer of ozone formed from oxygen in the higher atmosphere. Ozone is a very important gas; it blocks the ultraviolet radiation of the sun. Without ozone, we would be burned by this radiation and life at the surface of the earth would be impossible. In order for life to get closer to the surface, it was therefore necessary that the first algae, well-protected by thick layers of water, begin producing oxygen. We can understand why evolution is so complex: life evolves according to its surrounding environment and changes it at the same time.

Six hundred million years ago, when there was enough oxygen in the atmosphere, life started to explode. After that time, one finds a great variety of fossils of different organisms. This diversity has increased until today.

Successful Adaptations

What is truly marvelous in all the different living beings is their perfect adaptation to the environment.

The gull, for instance, is perfectly adapted to the

The expansion of life on dry land was possible only because algae enriched the atmosphere with oxygen. At very high altitudes, a protective layer of ozone was then formed.

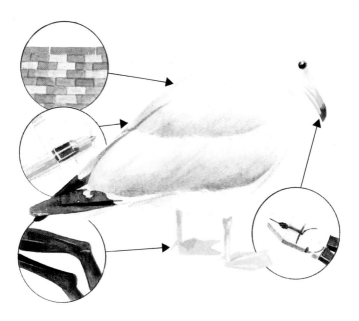

The sea gull is adapted to the marine environment *by the shape of its beak, the texture of its feathers, the web of its feet, and its sharp eyes. All these features allow the sea gull to live at the boundary between land and sea.*

marine environment. Each part of its body performs a useful function. Even the way it makes its nest expresses adaptation. The gull lays eggs on inaccessible cliffs to protect its offspring and associates with other gulls in colonies so as to be able to defend itself against possible predators. Other kinds of adaptations, which are not immediately obvious, are, for instance, special digestive juices.

Each animal is adapted to its environment, and every part of its body is utilized in that adaptation. The immense diversity of living forms is related to the immense diversity of their ways of life.

But how can the appearance of these adaptations be explained? How are new species, adapted to new ways of life, produced?

Also, how can the appearance of man be explained?

This is the problem of evolution, the subject of a completely different type of investigation—the topic of another book.

Diversity of the living world.

The Question Remains Alive

What a history! How many events from the Big Bang till today! How many complex constructions, encased one inside the other, from the atom to the living organism! All this resembles a giant puzzle. All the pieces are not yet in place, but large portions of the puzzle are recognizable. The general picture begins to emerge.

In fact, what do we know for sure?

The matter of living beings originated in the stars. The beginning of life is part of the general history of the universe. We are formed of stellar dust.

Step by step in the advancement of life, various constituents combined each time to form something more complex, richer in diversity and possibilities: a "chimera."

In this book, we have explored some scientific answers to the question, How did life on earth appear?

But *why* did life appear? Does life have a goal? These questions science cannot answer. It merely searches for the "how" of things. "How" and "why" are two completely different questions. One can explain scientifically "how" a car functions. But "why" I drive this car is no longer a scientific question; it is a goal that I freely choose for my trip.

In the future, science may perhaps be able to reconstruct a living being. But even though biology makes daily advances, nobody can make definite forecasts. If it ever happens, we shall understand better how life—this astonishing and marvelous thing that allows us to be here to wonder about the world—functions. But as to the question of "why," philosophy, religion, and—above all—each one of us must find the answer.

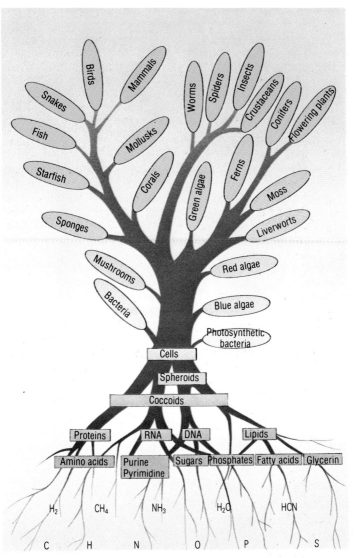

The tree of evolution. *Its roots converge toward the cell. Its branches diverge in the direction of the incredible diversity of the living world. This representation gives a simplified overall picture of evolution.*

73

GLOSSARY

Amino acids: Small molecules, the building blocks of proteins; all amino acids have a similar structure so that they are able to link together.

Astrophysicist: Physicist studying the structure and the evolution of the universe.

Atomic nucleus: Central part of an atom representing almost all of its mass and composed of protons and neutrons.

Atoms: Elementary particles that can combine to form molecules.

Bacteria: Microorganisms consisting of a single cell without a nucleus.

Big Bang: Great original "explosion" that started our universe.

Biologist: Scientist studying living beings, their structure, and function.

Cell: The basic living form. It is capable of self-preservation, self-regulation, reproduction, and can combine with other cells to form an organism.

Cellulose: Long chain of sugar molecules that forms the rigid walls of cells found in all plants.

Chlorophyll: Green light-trapping component found in the chloroplasts of plant cells. It is indispensable for photosynthesis.

Chloroplast: A chlorophyll-containing small organelle in a plant cell.

Clay: Earthy material, plastic when wet, consisting of weathered material from various rocks.

Coacervate: Microscopic structure formed by encasing, or nesting, of microspheres.

Coccoid: Very old microfossil (one billion years or more) that preceded bacteria.

DNA: Deoxyribonucleic acid. Long chain of nucleic bases that contains the genetic message.

Enzymatic protein: A specialized type of protein that controls chemical reactions inside cells.

Eucaryotic cell: Complex cell with a nucleus and numerous internal organelles.

Galaxy: Concentration of hundreds of billions of stars, gases, and dust particles. The Milky Way is one example of a galaxy.

Hydrothermal spring: Hot spring with very hot water that infiltrated close to the magma.

Interstellar cloud: Cloud of gases and dust particles spreading locally between stars.

Magma: Heated and molten rock inside the earth's crust. During volcanic eruptions, magma forms lava flows.

Meteorite: A rock traveling in space. Sometimes it collides with the earth's atmosphere and burns up as a shooting star.

Microsphere: Microscopic droplet enclosed in a membrane of proteinoids.

Mineral: Solid body at normal temperatures that, with other minerals, forms the rocks of the earth's crust.

Mitochondria: Internal organelles of a complex cell. They produce energy used by the cell.

Mutation: Random changes of DNA by addition, subtraction, or removal of nucleic bases.

Nuclear reaction: Bombardment of the nucleus of an atom by another atom or by a particle, thus forming a new atom.

Nucleic base: Complex molecule. Four slightly different types join to form chains in the DNA structure.

Organic molecules. Molecules found in all living beings. They almost always contain carbon atoms.

Ozone: Molecule composed of three atoms of oxygen. It is formed in the high atmosphere by the action of solar radiation that it absorbs.

Photosynthesis: Process that, with the help of solar energy, produces organic matter.

Planet: Heavenly body, relatively cold (thus without its own emission of light), gravitating around a star.

Predator: Animal feeding on animals of another species.

Proteinoid: Chain of amino acids produced in the laboratory and resembling proteins.

Proteins: Large complex chains of amino acids that form the essential part of living matter.

Radiation: Sum of radiations emitted by a body. A heated body radiates light.

Radioactivity: Particles and radiation emitted by a decaying nucleus of an atom.

Radioastronomy: Study of radio waves coming from heavenly objects.

RNA: Ribonucleic acid. Short chain of nucleic bases that participates in the transcription of DNA.

Solar system: The sun and all the planets, satellites, and asteroids that gravitate around it.

Species: A group of living beings that resemble each other and reproduce among themselves.

Spheroid: Fossil formed by a group of coccoids. It is the probable ancestor of the eucaryotic cell.

Star: Our sun is a star among others. Its radiation is produced by nuclear reactions.

Trilobite: Fossil animal that lived in great abundance 300 million years ago.

Ultraviolet radiation: Radiation located beyond visible violet light.

Virus: Small organic structure that can develop only inside a living cell.

Index

References to illustrations are in italics.

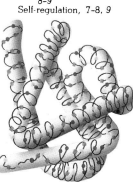

BARRON'S
FOCUS ON SCIENCE Series

Now young readers can feel all the excitement of scientific discovery! This series features lively texts plus lots of sparkling full-color photographs, drawings, maps, and diagrams. Each topic is clearly covered from its many angles, and all scientific concepts are thoroughly explained in simple terms. Each book is pocket size, and contains a handy index and a complete bibliography. *(Ages 13 and up)* Each book: Paperback, $4.95, Can. $6.95, 80 pp., 4¼" × 7⅛"

ISBN Prefix: 0-8120

THE ORIGIN OF LIFE
By Bernard Hagene and Charles Lenay

The many kinds of dinosaurs and other pre-civilization animals are explored here. Contains 53 full-color illustrations. (3836-3)

VOLCANOES
By Pierre Kohler

This colorful book explains what volcanoes are, how they're connected to earthquakes, and similar earth science topics. Contains 48 full-color illustrations. (3832-0)

DINOSAURS and Other Extinct Animals
By Gabriel Beaufay

This interesting book explains how our planet came into existence. Contains 58 full-color illustrations. (3841-X)

LIFE AND DEATH OF DINOSAURS
By Pascale Chenel

Here's a close-up look at how dinosaurs lived and how they became extinct. Contains 46 full-color illustrations. (3840-1)

PREHISTORY
By Jean-Jacques Barloy

The evolution of human beings is the focus of this exciting book. Contains 57 full-color illustrations. (5835-5)

All prices are in U.S. and Canadian dollars and subject to change without notice. At your bookseller, or order direct adding 10% postage (minimum charge $1.50), N.Y. residents add sales tax.

Barron's Educational Series, Inc.
250 Wireless Boulevard, Hauppauge, NY 11788
Call toll-free: 1-800-645-3476, in NY 1-800-257-5729
In Canada: 195 Allstate Parkway, Markham, Ontario L3R4T8

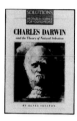